D0387863

DISCARD

Great Scientific
Questions and the
Scientists Who
Answered Them

SAN LEANDRO
LIBRARY
HIGH SCHOOL

HOW DO WE KNOW

THE LAWS
OF MOTION

JEREMY ROBERTS

Great Scientific
Questions and the
Scientists Who
Answered Them

HOW DO WE KNOW
THE LAWS
OF MOTION

THE ROSEN PUBLISHING GROUP, INC.
NEW YORK

For the motion poets and Mr. Frick

Published in 2001 by The Rosen Publishing Group, Inc.
29 East 21st Street, New York, NY 10010

Copyright © 2001 by Jeremy Roberts

First Edition

All rights reserved. No part of this book may be reproduced in any form without permission in writing from the publisher, except by a reviewer.

Library of Congress Cataloging-in-Publication Data

Roberts, Jeremy.
How do we know the laws of motion / Jeremy Roberts. — 1st ed.
p. cm. — (Great scientific questions and the scientists who answered them)
Includes bibliographical references and index.
ISBN 0-8239-3383-0
1. Dynamics—Juvenile literature. 2. Motion—Juvenile literature. [1. Dynamics. 2. Motion.] I. Title. II. Series.
QC133.5 .R63 2001
531'.1—dc21

00-012863

Cover images: The Delicate Arch, at Arches National Park in Utah, silhouetted against star trails.
Cover inset: pendulum.

Manufactured in the United States of America

Contents

Introduction

How do things move? It seems like such a basic question, with a simple answer. But it's not.

Imagine you're outside playing catch with a friend. You pick up a ball and throw it. He throws it back. The ball doesn't quite reach you, though. Instead, it bounces to the side, rolls down a hill, and changes direction when it collides with a telephone pole. It slowly rolls to a stop.

Why didn't it reach you? Why did it roll down the hill? Why did it change direction upon impact with another object, and why did it eventually stop? The answer to each question says something about motion. Things seem to move because a force is applied to them. One such force is gravity, and objects respond to gravity by moving toward the center of the earth. When one object collides with another, sometimes both objects will move off in the same direction, and sometimes they will move off in opposite directions.

The more you think about motion, the more complicated the subject becomes. Is all motion the same? If you walk backward on a train that is moving forward, in what direction are you really going? Earth spins on its axis at a speed of about 1,000 miles per hour. Why don't we fall off? What about objects in space, like the Sun and the Moon and the planets? No one seems to have thrown or pushed the Moon, yet it keeps moving every day and night in a regular pattern.

Understanding the behavior of objects in motion turned out to be one of the keys to understanding the universe itself. But it wasn't an easy task. Scientists first had to learn to rely on their senses. Then they had to learn how to disregard them. They developed powerful mathematical tools to describe motion and then discovered flaws in the theories that invalidated the mathematics. In the end, they had to relearn a lesson they had learned hundreds of years before—that our perception of motion can be influenced by how we observe it.

1

From the Ancients to Ptolemy

Humankind's earliest thoughts about motion are lost to us, but they were most likely very practical and concrete thoughts, rather than abstract ones. The first type of motion that was observed and recorded by ancient scholars was the motion of celestial bodies, such as the

Moon, the Sun, and the stars. Anyone who watches these heavenly bodies realizes that there are definite patterns to their movement. The movement of the Sun marks the length of a day. The Moon goes through a cycle of different phases as it moves, depending on how it catches the Sun's light, passing from a full Moon to a thin sliver and then to nothing in twenty-eight days. The stars turn in the sky every night, and a clever observer can use their movements to tell the seasons. The motions of celestial bodies helped the earliest civilizations to regulate the planting and harvesting of crops, and led to the creation of the first calendars. There was comfort and reassurance for people in the regularity of these celestial motions, and a sense of mystery that the movement of these bodies had something to do with the cycle of seasons and the renewal of life as winter changed to spring. So these celestial events played an important role in religious rituals that helped give societies order and meaning.

Different civilizations used different methods to study the motion of the heavens. Archaeologists believe that the people who built Stonehenge in England around 2000 BC erected the giant megaliths to help predict lunar eclipses and other astronomical events. At roughly the same time in Mesopotamia, the Babylonians were developing astronomy and mathematics to help them do the same thing. But while many early peoples carefully recorded such movements and could even make sophisticated predictions about them, they did not attempt what we might call a scientific analysis of why or how heavenly bodies moved. Their explanations for the movements of celestial objects were based on religious ideas or primitive analogies with the kind of motions they observed around themselves. The early Greeks, for

The Moon goes through a cycle of different phases during its twenty-eight-day orbit around Earth. When the Moon is becoming full, it is said to be waxing. Afterward, when it is decreasing in intensity, it is waning.

example, believed that the Sun was a god who rose out of the ocean and drove a chariot across the sky. When he reached the western horizon, he boarded a winged boat that brought him around again to the eastern horizon for the next day.

Telescopes weren't invented until the beginning of the seventeenth century, so all observations before that time were made with the naked eye. The movement of the Sun could also be measured by observing the movement of its shadow. By later standards, the observations and mathematical tools used by the Babylonians, Egyptians, and early Greeks were primitive. What today's scientists would consider a large margin of error was accepted as natural. Even so, the observations and calculations were reasonably reliable and useful in measuring earthly time.

The ancient people of the Americas paid close attention to the movement of the stars and planets, as is evident by the ruins of this Mayan observatory in Mexico.

THE EARLY GREEKS

In the sixth century BC, the people who lived in what is now Greece developed a tradition of science and philosophy that went far beyond anything that had come before. They made major advances in the natural sciences and in areas of mathematics such as geometry and trigonometry. Philosophers such as Thales of Miletus (624–546 BC) insisted on developing systems of thought that explained phenomena according to empirical or logical principles rather than religious ideas. Thales had been educated in "the east," that is, Egypt and Babylonia, and what he learned there about the motion of the planets enabled him to predict an eclipse of the Sun in 585 BC. The prediction astonished his contemporaries. Thales was also the first to use step-by-step proofs to demonstrate the truth of mathematical statements.

Greek philosopher Pythagoras (560–500 BC) practically turned geometry and numbers into a religion. Pythagoras is most famous for his study of triangles and the Pythagorean theorem (the sum of the squares of the lengths of the two shorter sides of a right triangle is equal to the square of the hypotenuse, or the longest side). But he was also the first observer to point out that the Sun, the Moon, and the other known planets moved differently from the stars and were probably not as far away.

ARISTOTLE

The Greek idea of motion—an idea that dominated European thinking for more than a thousand years—was developed by Athenian philosopher Aristotle (384–322 BC). Aristotle believed that Earth was at the center of the universe, and it did not move. How could it move? If it did, we would feel the movement and we would fall off. The Sun, Moon, and planets moved in perfect circles around Earth.

Given this scheme, Aristotle maintained that all objects have a tendency to move toward their "natural place," depending on what they are made of. The

Greeks believed that every object was made up of one or more of four basic elements: earth, water, air, and fire. If an object was made of earth, it tended to move downward toward the center of the universe. Water's natural place was just

Greek philosopher Aristotle though that the Sun, Moon, and planet. moved in circles around Earth

above the earth, just as the oceans covered the surface of the planet. If an object was made of air, it naturally moved upward. And if it was made of fire, its place was in the distant heavens. In this way, having no under-standing of the force of gravity (or what objects were really made of, for that matter), Aristotle tried to explain why a heavy lump of earth would fall from his hand to the ground without his applying any force to make it move. The speed of objects toward or away from the earth depended on their weight. Heavier objects fell to the ground faster than lighter objects, or so Aristotle thought.

Aristotle also observed that any object set in motion, on the earth at any rate, will eventually slow and stop. He believed that this was a natural tendency of things. Having no understanding of the concept of friction, he believed that it was natural for objects to return to a state of rest. In order to keep an object moving, it was necessary to apply a force to it continuously. Something was continually pushing or pulling the planets to keep them moving in their orbits. In Christian times, it was thought that angels pushed the planets and made the universe turn about Earth.

Aristotle's ideas about motion were based partly on what seemed to him and others to be common sense. They were not derived from extensive experiments or the systematic observation of motion as a phenomenon. And yet Aristotle's ideas became the framework that philosophers and scientists would use for centuries. Aristotle's ideas were spread first at the Lyceum, a school he founded in Athens, and then throughout the known world. Not all of his students or later commentators agreed with everything he said. But for generations Aristotle's work represented the foundation of what passed for science. When in the Middle Ages the Catholic Church endorsed Aristotle's teachings, it became heresy to question his ideas.

The problem was that when observers actually began to study the motions of the planets, those motions did not conform to the theory. According to Pythagoras, Aristotle, and the Greeks in general, the orbits of the planets had to be perfect circles, and the planets had to move in their orbits at a constant speed, in a constant direction, at a constant distance from Earth in the center of things. Even with the primitive observational

Despite the fact that he was wrong, Ptolemy explained the motions of the planets and predicted their future positions.

tools available at the time, it was obvious to those who looked that the planets were sometimes closer to Earth than at other times, and that they moved in their orbits at varying speeds. Indeed, sometimes a planet appeared to slow down and double back in the opposite direction!

PTOLEMY

Elaborate efforts were made to make the facts fit the theory. In the second century AD, Greek astronomer Claudius Ptolemy of Alexandria (100–170) adopted and

21

popularized an astronomical system developed by Hipparchus of Nicea (190–120 BC) about 200 years earlier. In this system, Earth was still at the center of the universe and the orbits of the planets were still perfect circles, but the planets were given additional motions. In addition to moving in their orbits, they were also revolving in mini-orbits called epicycles. These epicycles could explain the "retrograde," or backward motion, of the planets. There were other complications, and the orbits of the planets were now pictured as a complex system of wheels within wheels, like the inner mechanism of an old mechanical watch. Yet for all its awkwardness, obvious to us today, the Ptolemaic system did explain the curious motions of the planets and allowed for the prediction of their future positions. And it did not challenge the Aristotelian idea that an

Ptolemy adopted and popularized Hipparchus's astronomical system. It maintained that Earth was at the center of the universe, and the planets orbited it in circles and mini-orbits, called epicycles.

unmoving Earth stood at the center of the universe and all other celestial bodies revolved around it.

After the collapse of the Roman Empire in the fourth century, active scientific inquiry ceased in Europe for almost a thousand years. But Ptolemy's ideas were preserved in Arabic writings and translated back into Latin in the twelfth century. His theories were adopted by the Catholic Church, which found that an Earth located at the center of the universe fit well with religious beliefs and the notion that human beings were the Lord's favored creatures. In the Catholic Church's fierce struggles against heresy in the centuries to come, it became dangerous to suggest that Ptolemy might be wrong. Ancient thinkers like Ptolemy and Aristotle were regarded by the Catholic Church as "authorities" who, though they lived in the pre-Christian era, had put forward doctrines that contained divine wisdom. Scholars who questioned these authorities might have publication of their books prohibited, or they might even be imprisoned or executed. To push the boundaries of science beyond the ancient Greeks was going to require courage.

2

Copernicus, Kepler, and Galileo

Though it proved useful in predicting the motions of the planets, over the short term at any rate, Ptolemy's theory was amazingly complex, and many philosophers and scientists were disturbed by it. It made the solar system look jury-rigged, like a clunky machine fitted out with additional gadgets to make it work properly. The model lacked the elegance and simplicity that even Christian scholars thought God's work should reveal. In the early fourteenth century, the Franciscan monk and Oxford scholar William of Ockham (1285–1349) put forward what would become an important scientific principle. He said that "entities must not needlessly be multiplied." By this he meant that when competing theories about a phenomenon are considered, the one that makes the fewest and simplest assumptions is probably correct. This principle is known as Ockham's razor because it tries to cut away unnecessary complications. For those scholars who found this idea compelling, Ptolemy's model of how the planets moved just didn't seem right. What's more, if one was very bold in one's thinking, there was a simpler explanation.

In 1496, Polish astronomer Nicolaus Copernicus (1473–1543) traveled to Italy to attend a conference on reform of the calendar. Calculations of the planets'

Fearing charges of heresy from the Catholic Church, Copernicus delayed publishing his theory of the solar system.

future positions using Ptolemy's system were not producing results that agreed with the calendars in use at the time, and scholars knew that something had to be done. The mood in Renaissance Italy was one in which scholars might question ideas, and this is what Copernicus did. In a bold leap of imagination, he suggested a model for the solar system that placed the Sun at the center of the universe, with all the planets revolving around the Sun. His model immediately solved many problems. The retrograde, or backward movement, of the planets Mars, Jupiter, and Saturn occurred because they were in larger orbits farther away from the Sun than Earth, and Earth in its

orbit would occasionally swing ahead of these planets, making them seem to move backward.

Copernicus delayed publishing his theory for many years because he feared the reaction of the Catholic Church, but his ideas finally appeared in a book in 1543, the year of his death. It is not known if he ever saw the book before his death. Intended for circulation among a select group of scholars, only a few hundred copies were printed, but his ideas began to gain followers. Unfortunately, while his theory placed the Sun at the center of the universe, it did not challenge the other Ptolemaic ideas of how planets moved, at constant speeds and in perfect circles. To explain this, Copernicus had to retain some of Ptolemy's complicated epicycles, and the simplicity of his new theory was marred.

The Copernican model of the solar system postulated that the Sun was at the center of the universe and the planets orbited around it.

ELLIPTICAL ORBITS

The problem was finally resolved by German astronomer Johannes Kepler (1571–1630). By the time Kepler graduated from the University of Tubingen, he had adopted Copernicus's ideas. In 1598, he accepted a teaching position at the University of Prague, where he met and befriended the great Danish astronomer Tycho Brahe (1546–1601). Kepler's eyes had been damaged by an attack of smallpox when he was three years old, and he was never very good at making astronomical observations, but this was Tycho's strength. When Tycho died in 1601, Kepler inherited all his observational data, including very complete observations on the motion of the planet Mars.

This illustration depicts Tycho Brahe working in his observatory, where he recorded many astronomical observations.

COPERNICUS, KEPLER, AND GALILEO

Kepler studied Tycho's data and realized something that no one else had realized: that Mars and the other planets moved in elliptical, not circular, orbits. An ellipse is a flattened circle. The degree of flatness is a measure of its "eccentricity." Planetary orbits were not very eccentric. They were almost circles, but not quite. Kepler worked out a number of mathematical laws that described the motion of planets in elliptical orbits and not moving at constant speeds. The closer they came toward the Sun, the faster they moved. This suggested to Kepler that somehow the Sun exerted a force that controlled the movement of the planets. But beyond suggesting that this force might be magnetism, Kepler left that problem unresolved.

By introducing the idea of elliptical orbits, and by dismissing the notion of perfect circles and uniform

Kepler proved that the positions and movements of the planets could be calculated accurately, if it was assumed that their orbits were elliptical.

orbital speed, Kepler had provided the mathematical foundation for the acceptance of the Copernican theory of the solar system and swept away a thousand years of confusion. Now the positions and movements of the planets could be calculated with a high degree of accuracy.

Every scientific advance, however, raises new questions, and the Copernican theory was no exception. Though Kepler may have accurately explained the motions of the planets, objections were raised because the Copernican theory seemed to lead to an absurdity. If Copernicus was right, Earth was moving through space at incredible speeds as it revolved around the Sun. Furthermore, to explain the motions of the stars, it was now necessary to suppose that Earth was spinning on its axis. If Earth was moving in this fashion, why don't we fall off? Why don't we at least feel like we are moving?

It was a question that neither Copernicus nor Kepler could answer, but Kepler knew the man who would answer it. From Prague, he had been exchanging letters with another disciple of the Copernican theory, a rather brash and outspoken professor of mathematics working at Padua near Venice in Italy. His name was Galileo Galilei (1564–1642).

THE RELATIVITY OF MOTION

Galileo was an extremely popular lecturer in the sciences, and he enthralled his students by demanding that they test their ideas through experimentation and empirical observation rather than trusting the authority of the ancients. This Galileo did himself when he became interested in the motion of falling bodies. Aristotle had said that the heavier an object, the faster it would fall.

Galileo wanted to test this idea. Legend has it that he dropped two objects of different weights from the top of the Leaning Tower of Pisa, but this is not true. The objects would have fallen much too fast for

him to measure the difference in their times of fall. There were no stopwatches in Galileo's time. Indeed, there were no clocks at all. All he had to measure time was his own pulse. The experiment was actually carried out by "slowing down" gravity, by rolling objects of different weights down inclined planes. Galileo discovered that all objects near the surface of the earth fall at the same rate of acceleration, regardless of their weight. In proving this point, he overthrew the authority of ancient Greek science. Galileo's work on falling bodies laid the groundwork for much of our understanding of gravity as a force. But he didn't quite develop a theory of gravity. That objects were attracted to the earth was just an assumption he made, along with all the thinkers who preceded him.

Observing the movement of cannonballs, Galileo realized that without the force of gravity pulling the cannonball to the ground, the ball would travel in a straight line forever. This also applies to planets and stars.

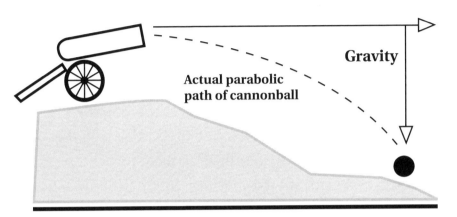

**Motion imparted
by cannon**

Gravity

Actual parabolic
path of cannonball

Angular momentum of planet

Actual orbit of
the planet

Earth

Gravity

Sun

Galileo also studied the motion of cannonballs and observed that objects could be subject to more than one force at a time. The force imparted to the cannonball by the explosion of the cannon made the ball travel in a straight line, but the same force that made objects fall to the earth also made the cannonball travel downward, so that the total motion was a parabola. Galileo concluded that without the force pulling the cannonball to the ground, the ball would travel in a straight line forever. This contradicted Aristotle as well, who believed that objects had to be continually pushed or pulled to move. Galileo was saying that, after the initial impulse, an object might move forever until opposed by another force. If force were continually applied to an object as Aristotle required, an object would accelerate. It would continue to go faster and faster, like a falling object. This immediately simplified the whole notion of how planets moved in their orbits. God might provide the initial impulse to start the planets moving, but no other explanation was required to explain why they continued to move. They moved simply because no force stopped them from moving.

It was his defense of Copernican theory, however, that led to a revolutionary idea in the study of motion. In order for the Copernican theory to work, Galileo had to answer the critics who held that if Earth was really moving through space we would all be thrown off the planet. At the very least, we would feel Earth's motion. But Galileo knew better. He used the example of sailors on a ship. When the ship's motion was not constant, when it turned, speeded up, slowed down, or rocked with the waves, certainly sailors could feel the ship's motion. They might, in fact, have trouble standing upright on a violently tossing ship. But if the ship was not turning or changing speed, and if it were moving at a constant speed on a calm sea, the sailors could walk about the deck as easily as they could walk on land, and they would not feel the motion of the ship. If they closed

their eyes, they might not even be able to tell which direction they were moving in, or if they were moving at all. As long as they were moving with the ship and the ship's motion was constant and uniform, that motion could not be detected.

It followed that if Earth was moving through space at a constant and uniform speed, the people on Earth would not be affected by that motion and would not even be able to sense it because they shared the same motion with Earth. As scientists would express it today, people are part of the same "inertial frame of reference" as Earth because they are moving with it through space. In order to detect uniform motion, an observer has to be part of a different frame of refer-ence. A person on another planet, for example, would observe the movement of Earth relative to his or her position. People on Earth could infer the motion of the earth by watching how other planets moved rela-tive to us. All of us have the example of closing our eyes on a moving train or in a moving car and losing

In 1633, the Catholic Church tried Galileo for heresy for his "radical" ideas. Even though Galileo was later proven correct, it wasn't until 1992—some 350 years later—that the church acknowledged it was in error for condemning him.

track of which way we are traveling. This is because you are in the same frame of reference as the train or car. You have to open your eyes and observe the scenery passing relative to the train or the car in order to be certain of the direction of movement. Steady, uniform motion can be detected only relative to something outside the moving system.

Galileo published his radical ideas in a book entitled *Dialogue on the Two Chief World Systems* in 1632, and the next year he was called before the pope and tried for heresy. Just a generation earlier, in 1600, Italian philosopher Giordano Bruno had been burned at the stake for equally heretical ideas, so Galileo publicly repudiated his own theories, though privately he was certain they were correct.

The trial, however, could no longer halt the progress of science. Galileo's book had been published, and his ideas, as well as those of Kepler and Copernicus, were spreading across Europe. In 1642, the year that Galileo died, Isaac Newton (1642–1727) was born in Woolsthorpe, England. Newton would take up and expand upon Galileo's concepts of motion, and the result was a revolution in our understanding of the universe.

3

The Laws of Motion

As a young man, Isaac Newton did not appear to be a particularly brilliant student, but he did show an interest in mechanical devices. With the help of an uncle, he was admitted to Cambridge University, and he graduated in 1665. He was in London when the plague began to strike the city,

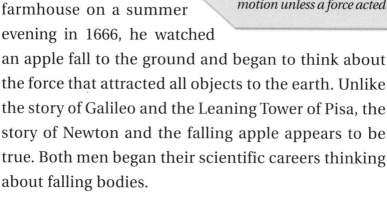

and to escape the danger he retired to his mother's farm in Woolsthorpe, in Lincolnshire, where he had been born. He was twenty-four years old. Sitting behind the farmhouse on a summer evening in 1666, he watched an apple fall to the ground and began to think about the force that attracted all objects to the earth. Unlike the story of Galileo and the Leaning Tower of Pisa, the story of Newton and the falling apple appears to be true. Both men began their scientific careers thinking about falling bodies.

Isaac Newton realized that a body at rest or in motion would remain at rest or in motion unless a force acted upon it.

Newton lived at a time when modern science was finally emerging and separating itself from ancient beliefs. The accepted wisdom had been that the heavens and the earth were governed by two separate sets of natural laws. But Newton wondered if the same force that made the apple fall to the ground also held the Moon in its orbit around Earth. Just as a cannonball traveled in a parabolic path, moving laterally but eventually falling to the earth, it was possible to view the Moon's orbital motion as a continuous fall toward Earth. Newton believed that the strength of this gravitational force would weaken rapidly as the distance between objects feeling the force increased.

Using a new type of mathematics he was developing, known as calculus, Newton tried to determine the strength of the force acting on the Moon and to see if it bore any relation to the way falling objects accelerated on the earth. Unfortunately, something about the calculation didn't make sense. He may not

have had a correct measurement for the distance between Earth and the Moon, or for the radius of Earth, or there may have been some other problem. Discouraged, he abandoned the problem for fifteen years. In the interim, he became famous for using a prism to prove that sunlight was made up of a spectrum of different colors. He perfected his calculus, and he invented the reflecting telescope.

By 1669, Newton found himself back at Cambridge, now the prestigious Lucasian professor of mathematics, and he remained at Cambridge for thirty years. In 1684, architect Christopher Wren offered a prize to anyone who could work out the laws governing the motion of the planets. Newton's friend, astronomer Edmund Halley, urged him to try again. With more accurate data, this time Newton succeeded. Yes, the

Isaac Newton is depicted here with his new reflecting telescope.

apple falling to the earth and the motion of the Moon could both be explained by one universal law of gravitation. That force was directly proportional to the masses of the two objects attracting each other, and inversely proportional to the square of the distance between the objects. That is, the more massive two objects, the greater the force of attraction between them, but the farther away they were from each other, the weaker the force. Newton represented the relationship with the following equation:

$$F = G \ \frac{mm_1}{D}$$

where F was the force of gravity, m and m_1 were the masses of the two objects, D was the distance between them, and G was a constant, a number that was not actually discovered until a century later. Furthermore, using his calculus, Newton determined that this formula would produce elliptical orbits for the planets, confirming Kepler's work. The universal law of gravitation expressed in this formula explained the motions of all the planets according to one measurable force. It revealed that the solar system worked very much like a huge machine that could run without the intervention of supernatural forces, and this was an intellectual revolution of enormous importance.

The formula for the first time explained what Galileo had discovered: that all objects of whatever mass fall to the earth at the same speed. Yes, the more massive an object was, the more gravitational force it felt, but the more massive it was, the more it also resisted the force of gravity. The property of mass that resists displacement by a force is known as "inertia." The balance between gravity and inertia resulted in every object near the surface of the earth falling at the same rate of acceleration, 32 feet, or 9.8 meters, per second per second.

In 1687, Newton published his findings in his *Philosophiae Naturalis Principia Mathematica*, "Mathematical Principles of Natural Philosophy," still widely regarded as the greatest work of science ever written. The first edition was written in Latin and ran to no more than 2,500 copies, so its impact was felt only by a handful of educated people. Yet its date of publication is considered the beginning of the Age of Reason. Newton received many honors as a result and died a wealthy man, buried in Westminster Abbey. The poet Alexander Pope wrote:

> *Nature and Nature's laws lay hid in night:*
> *God said, Let Newton be! And all was light.*

Newton himself said:

> *If I have seen further than other men, it is because I have stood on the shoulders of giants.*

Principia Mathematica did much more than state Newton's universal law of gravitation. In the book Newton had reexamined and reorganized all of Galileo's work on the motion of bodies into three fundamental laws of motion that would guide scientists for the next 400 years.

Newton's first law was the law of inertia. It stated that a body at rest or in uniform, unaccelerated motion would remain at rest or in uniform motion unless a force acted upon it. That is to say, objects resisted being pushed or pulled by forces. Newton had not been fooled, as Aristotle had, by the force of frictional resistance that slowed objects down in most earthly situations. He saw that friction was simply another force, not a property of objects that required them to be continually pushed or pulled in order to maintain their speed. If objects were moving, it was because a previous force had acted on them, and no further explanation was required for their movement. They would continue moving forever until some other force intervened.

Newton's second law defined the measurement of force as the mass of an object times its acceleration, and is expressed in the formula:

$$F = ma$$

where F is force, m is mass, and a is acceleration. That is, how much the speed or direction of an object of a certain mass, at rest or in uniform motion, is changed is directly proportional to the amount of force applied to it. In the metric system, for example, if an object at rest with a

mass of one kilogram (about 2.2 pounds) were accelerated at a rate of one meter (about three feet) per second
per second, it would have been subjected to a force—a
push or a pull—of one "newton," so named after the
great scientist. One newton, incidentally, is equivalent
to the force of gravity acting on a quarter-pound bar of
butter. How fast this object would go would depend on
how long that force was applied. If it were applied for
five seconds, the object would have a speed of five
meters per second. If two newtons of force were
applied, the one kilogram object would accelerate at
two meters per second per second, or twice as fast.

Newton's third law is most familiar to anyone
who has studied rocket propulsion. It states that for
every action there is an opposite and equal reaction.

*The Saturn rocket with the Apollo 11 spacecraft is launched to the
Moon. Its flight path was calculated according to Newton's laws.*

To propel a rocket of a certain mass forward at a certain speed, exhaust gases must be expelled from that rocket in the opposite direction. The mass of those gases times their acceleration backward out of the rocket engine must equal the mass of the rocket times its forward acceleration.

With the universal law of gravitation and the three laws of motion laid out by Newton in *Principia Mathematica,* scientists were given simple and clear rules for calculating the motion of objects and the magnitude of the forces acting upon them. It seemed that virtually all the unexplained aspects of planetary motion were now explained, and the cosmos was seen to work like a kind of huge machine, an intricate collection of objects governed by measurable forces. Newton's universe has been called the clockwork universe for this reason.

What a "force" was, however, was not so clear. Newton defined it as an "action at a distance," which was both profound and vague. It seemed easy enough to understand how force was transmitted when two

The steam engine, pictured here at the 1876 American Centennial Exhibition in Philadelphia, was a major technological achievement. It came to symbolize a mechanical world governed by Newton's laws of force and motion.

objects collided with each other, when they made actual contact, but the gravitational force was not like that. It exerted its influence over great distances, through a vacuum, on objects that were not connected in any way.

Newton's laws provided the basis for a great many scientific and engineering achievements over the next several hundred years. They enabled people to work out all sorts of practical problems, such as how much strength a bridge would need to support itself and its load, or how far a projectile would travel when shot from a cannon. The eighteenth and nineteenth centuries saw immense advances in the application of machine power to industry, at least partly because of Newton and the scientists who followed him. Newton's laws are still taught and used today, and in most situations they produce useful results. These laws are even used to plot the flight paths of spacecraft. The laws of motion represented an enormous and revolutionary advance for the scientific view of the world.

Some time in the late nineteenth century, however, a few very bold scientific thinkers began to realize that these laws could not completely describe the motion of bodies.

4

It's All Relative

In 1871, Scottish physicist and mathematician James Clerk Maxwell (1831–1879) arrived at Cambridge University, a newly appointed professor of experimental physics. Maxwell was a brilliant thinker who already had made several important contributions to science. He had

proved that the rings of Saturn had to be composed of numerous small solid rocks. He had done work with gases that lent support to the kinetic theory of matter, the theory that temperature and pressure were caused by the movement of molecules. But now, for the last few years, he had become interested in the work that Michael Faraday (1791–1867) had done with electricity and magnetism a generation earlier.

Faraday had conducted numerous experiments that demonstrated a relationship between electricity and magnetism. A wire circuit moving through a magnetic field would generate an electric current, and a flowing electric current seemed to create a magnetic field. Electric and magnetic fields always occurred together, and Faraday had used this fact to construct

The Faradays discovered that a wire circuit moving through a magnetic field generates an electric current, and a flowing electric current creates a magnetic field. Michael Faraday used this knowledge to build the first generator.

the first trans-
former and the
first generator.
Faraday, however,
lacked the math-
ematical knowl-
edge to do more
than note this
relationship
and conduct
practical experiments
with it. Now it was Max-
well's turn. If Maxwell was
good at anything, it was mathematics.

James Clerk Maxwell described the relationship between electric and magnetic fields.

Maxwell developed a few simple, and now famous, equations that describe the relationship between electric and magnetic fields and that demonstrated that the two phenomena were really one force, electromagnetism. The equations also revealed that electromagnetic fields took the form of waves that radiated outward from their source at the speed of about 300,000 kilometers, or 186,000 miles, per second. As Maxwell knew, this was also the speed of light. This was too much of a coincidence, and Maxwell theorized that light was actually a form of electromagnetic radiation.

Waves, of course, need a medium to convey them, just as sound waves are carried by air molecules and ocean waves are formed from water molecules. Waves are not things in themselves. They are impulses of energy passing through a material medium. Electromagnetic waves, however, seemed able to pass through a vacuum. Light waves could move through the vacuum of space. To explain this, Maxwell reverted to an idea that went all the way back to Aristotle. He said that all of space must be permeated by an invisible and undetectable "luminiferous [light-carrying] ether." Light waves were in fact the oscillations of this ether. This invisible ether was pretty strange stuff, but if it permeated all of space and it could carry forces, it might explain Newton's idea of force as "action at a distance."

MICHELSON AND MORLEY

From 1881 to 1887, first in Berlin and then at the University of Chicago in collaboration with Edward Morley (1838–1923), physicist Albert Michelson (1852–1931) conducted a series of increasingly sophisticated experiments designed to detect the ether.

Michelson designed an instrument called an interferometer. The final version consisted of a huge, turnable stone wheel, upon which were a light source and a complicated set of mirrors that could split a beam of light into two beams, direct them at right angles to each other, and then rejoin them at the observer's position. Michelson reasoned that if Earth was moving through space in its orbit around the Sun, Earth must

Albert Michelson (seen here) and Edward Morley discovered that light travels at the same speed regardless of its orientation in space. In doing so, they also managed to debunk the existence of ether.

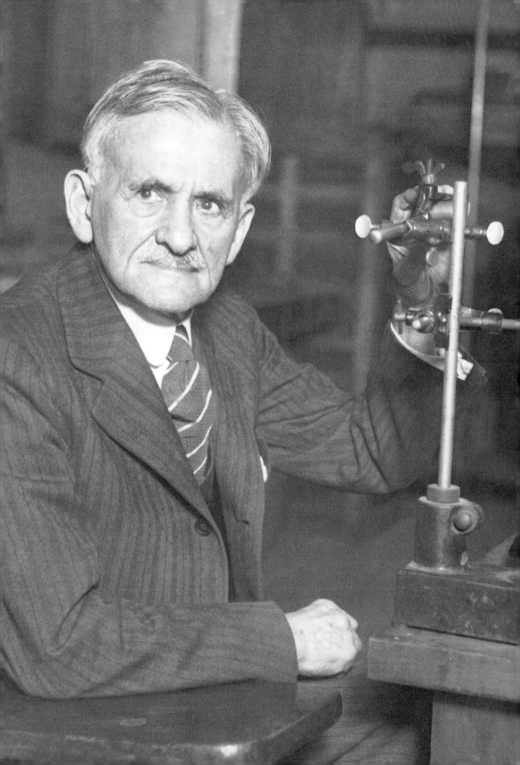

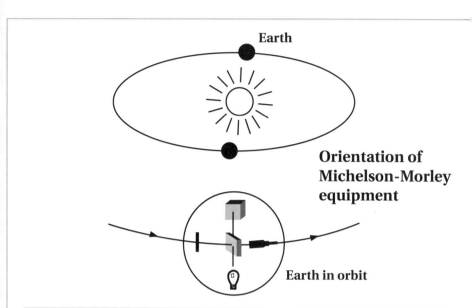

Earth

**Orientation of
Michelson-Morley
equipment**

Earth in orbit

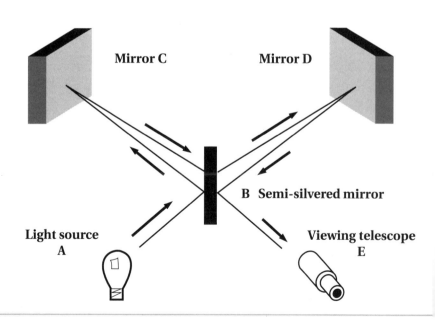

Mirror C   Mirror D

B Semi-silvered mirror

Light source
A

Viewing telescope
E

be moving through the ether. If the split light beam could be aligned so that one beam traveled parallel to Earth's movement while the other beam moved perpendicular to that movement, the two beams would travel at slightly different speeds. When they were rejoined at the observer's position, they would be "out of phase." The observer would see a fuzzy image with interference rings around it, rather than a single, well-defined image. That would be proof that the ether existed because it had accelerated the beam of light moving in the same direction as the ether, just as a boat in a river goes by the shore faster if it is moving downstream.

The ether, incidentally, because it permeated all of space but allowed material objects to move through it, constituted a kind of invisible framework that

Michelson's interferometer was a turnable stone wheel with a light source and a set of mirrors to split a beam of light into two beams, direct them at right angles to each other, and then rejoin them at the observer's position.

defined the dimensions of the universe. If one could detect motion through the ether, that meant that one could detect "absolute motion" in relation to the boundaries of the cosmos.

No matter how often they tried, however, Michelson and Morley could not detect the expected interference patterns, and they were forced to conclude that the beam of light traveled at the same speed regardless of its orientation in space. There was no evidence that the ether existed. Absolute motion could not be measured.

Here's an analogy that demonstrates just how odd these results were. Imagine that you are driving on a highway at a certain speed, and your brother, traveling alongside you in another car, is keeping pace with you. You know that both of you will arrive at your destination at the same time, as long as you both travel at the same speed and go the same distance. But you decide to step on the gas and accelerate, while your brother plods along as before. Imagine how surprised you would be if, having floored the gas pedal, you watch as your brother still arrives at your destination exactly when you do. Newton's second law of motion clearly states that the more force you apply to an object, the faster it accelerates. But if both vehicles were traveling at the speed of light, no matter how much more force you applied, you could not go faster than the other vehicle.

SPECIAL RELATIVITY

It made no sense at all that a beam of light traveled at the same speed regardless of its orientation or point of origin. If, for example, we observed the light from two objects, one moving rapidly toward us and one moving rapidly away from us, Newtonian physics held that there ought to be a great difference in the speed of those two light beams. The beam from the object moving rapidly toward us ought to arrive at our position much more quickly than the other beam. Its speed ought to be the sum of the speed of light plus the speed

of the approaching object. The speed of the other beam of light ought to be the speed of light minus the speed of the receding object. But the Michelson-Morley experiment and other physical experiments demonstrated that this was not the case. No matter what the circumstances, measurements of the speed of light always produced the same result.

Ironically, the man who would explain all this, Albert Einstein (1879–1955), claimed to have been unaware of Michelson's experiment and its astonishing result. Born in Germany during the rise of Prussian militarism, Einstein, a pacifist, immigrated to Switzerland to escape service in the kaiser's army. He obtained his college education in Zurich and then went to work as a clerk in the Swiss patent office. This was not an academic position, and Einstein may not have had easy access to the scientific journals that reported on Michelson's work. But the job gave him plenty of time to think about scientific questions, and after years of rote learning in a German high school, he had developed a rebellious attitude toward accepted wisdom.

Instead of puzzling over Michelson's interferometer experiment, Einstein puzzled over Maxwell's equations describing electromagnetism. They contained the same contradiction. Electromagnetic waves such as light waves had to propagate from their source at the speed of light. But what if an observer could accelerate himself to the speed of light and travel parallel to the electromagnetic wave? Would the light wave appear stationary to the observer? Would the light wave cease to wave? Then, for the observer, it couldn't exist. No, the equations implied that no matter how fast the observer went, the light wave would still appear to shoot away from him at the speed of light. This "thought experiment," as Einstein called it, led to the same conclusion as Michelson had reached. The speed of light in a vacuum was a universal constant and never varied.

If the speed of light was an absolute, Einstein reasoned, then our concepts of space and time were

incorrect. Speed was defined by a simple formula, the amount of distance covered divided by the time taken to cover that distance. To determine a speed, you measured a distance traveled in a certain time.

One of the greatest geniuses of all time, Albert Einstein formulated the special theory of relativity.

If the answer (the speed of light) never varied regardless of the distance traveled, then time had to vary. Numerous scientific analogies could demonstrate this. The one Einstein chose involved events on a rapidly moving train observed from two points of view.

Suppose you are standing on an embankment by the side of a railroad track watching a very fast moving train roll past you. You can observe what is going on inside one of the cars, where a person has set up a simple device. On the ceiling of the railroad car is a flashing lantern that also triggers a stopwatch when it flashes, and on the floor of the car is a light-detecting device that stops the stopwatch when a beam of light from the lantern above hits it. The stopwatch will record the amount of time it takes the beam of light to travel from the ceiling of the railroad car to the floor, and the person on the train will record this time interval. You, too, standing outside the train on the embankment, have a stopwatch, and it is presumed that you can also accurately record the time it takes the beam of light to reach the floor of the car. Everyone is ready. The train is rolling forward. The lantern on the ceiling of the car flashes on.

To the observer in the railroad car, the experiment produces a simple result. The beam of light travels straight down from the lantern on the ceiling to the detector on the floor and trips the stopwatch. The

Train traveling at high speed

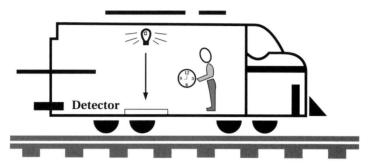

Detector

The way things appear to a person inside the train

The way things appear to a person outside the train

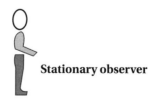

Stationary observer

observer records that this took a certain amount of time. For you on the embankment, however, this measurement is not so simple. From the embankment, you see the lantern in the car flash on, but before you can record the arrival of the beam of light at the detector on the floor, the whole railroad car has moved forward a certain distance. Light travels very fast, of course, and this change in the railroad car's position would not be very significant if the train was traveling at normal speeds. But remember that we have speeded up this train to a significant portion of the speed of light. The train will move down the track a certain distance before the beam of light can complete its trip.

From your point of view on the embankment, the distance between the lantern on the ceiling of the car,

Motion depends on your perspective or frame of reference. If you are on the train, the light appears to move vertically. If you are watching from outside the train, the light appears to move diagonally.

where the light beam originated, and the new position of the detector on the floor of the car is a diagonal line, a longer distance than that observed by the person on the train. Since the speed of light is constant, it will take the light beam a longer time to move this diagonal distance than the vertical distance observed by the person on the train. Your stopwatch will actually record a time of travel for this light beam that is different from the time recorded on the train.

If we speeded up the train even more, to you on the embankment this diagonal line would grow even longer, though to the observer in the railroad car nothing would change. The time it takes the light beam to traverse this diagonal would grow longer and longer as the train went faster and faster. If the train were able to achieve the speed of light itself, the beam of light would move horizontally and would never reach the detector on the floor. It would take an infinite amount of time to reach the detector. The faster the train moves, the greater this effect of measuring a lengthening of time, and the greater the difference in time between observers in different frames of reference.

Einstein called this the "relativity of simultaneity." It was the principle that observers in different frames of reference would not measure events as occurring simultaneously. This was true regardless of how fast objects moved in relation to each other, but the difference in measurement was insignificant at most speeds we are familiar with, and Newton's mathematics worked well enough under these conditions. But with objects that moved close to the speed of light, observers not moving with those objects would measure significantly different time intervals from those measured by observers on or in those objects. Since the universe was a collection of objects all moving at different speeds in relation to each other, it followed that there was no way to measure absolute motion.

Einstein developed new equations to describe motion at or near the speed of light, drawing on the

mathematical work of George FitzGerald (1851–1901) and Hendrik Lorentz (1853–1928), both of whom had come up with the same formula to explain the results of the Michelson-Morley experiment. According to these new equations, it wasn't just time that appeared to slow down as you observed an object approaching the speed of light. An object would also appear to grow shorter in length in the direction it moved as it approached the speed of light, until at the speed of light the equation stated that the object would shrink to zero thickness. An object would also grow more massive as it approached the speed of light. This did not mean that it got bigger or weighed more. It just came to possess greater and greater inertia, or resistance to a change in its motion, and required increasing amounts of force to push it even faster. At the speed of light, it would take an infinite amount of force to push an object faster. This made it quite plain that a speed just short of the speed of light was the maximum speed any material object in the universe could attain.

All of these changes in time, length, and mass that occurred as objects approached the speed of light were detectable only by observers looking in from outside, so to speak, by observers in a different inertial frame of reference. If you were moving with the object, like the observer in the train, you would detect no abnormality in length, mass, or time. This paralleled Galileo's idea that you could detect uniform motion only from the perspective of some-one not part of the moving system. For this reason, Einstein called his theory the "theory of relativity."

The implications of the theory were enormous. The speed of light was the only absolute in the universe, and space and time were not fixed but variable quantities, depending literally on one's point of view. And as one approached the speed of light, Newtonian mechanics could no longer explain the relationship between force and mass and acceleration. At the speed of light, the application of more force did not mean more acceleration, but more resistance to

acceleration. These ideas were so extraordinary that many scientists had trouble accepting them.

When he first published his results in a short scientific paper in 1905, Einstein referred to his discovery as the special theory of relativity because it dealt only with uniform motion in a straight line. He knew that there was another kind of motion that required a different explanation, but it would take him another ten years to puzzle out the answer to that question, and the result would be a revolutionary new understanding of the universe.

The Geometry of Space

The special theory of relativity had demonstrated that space and time were not absolute quantities but were "flexible," so to speak, and measurements of space and time varied depending on the motion of the observer. This shaking up of conventional notions of space and time freed

Einstein for another great leap of imagination as he tried to generalize his theory to account for all forms of motion. The problem Einstein set himself to now was to explain accelerated motion.

Here we have to introduce a distinction between speed and velocity. Speed is simply a measure of how much distance an object covers in a given time, regardless of the direction of movement. But velocity is defined by two quantities, both speed and direction of movement. Accelerated motion involves a change in either of these quantities. If an object changes its speed, either by speeding up or slowing down, it is accelerating. If an object maintains the same speed, but changes its direction of movement, it is also accelerating. Any change in velocity is a form of acceleration.

Even Galileo knew that his ideas about the relativity of motion applied only to uniform, unaccelerated motion. Uniform, unaccelerated motion could be detected only by someone in a different frame of reference. You could see a car drive past you on a road and

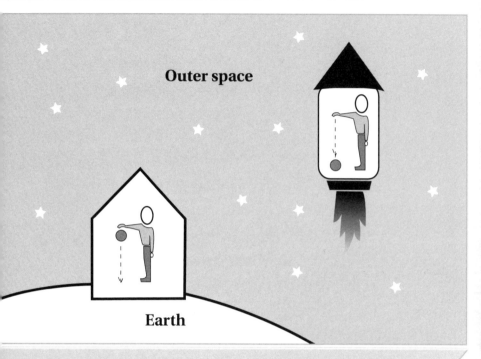

Outer space

Earth

Without an observation from another frame of reference, you cannot tell the difference between acceleration and gravity. Einstein called this the principle of equivalence.

measure its speed. But if you were in the car and the road was very smooth and your eyes were closed, you might not be able to detect your motion, and certainly you would be confused about which direction you were going in.

This was not true, however, for accelerated motion. Accelerated motion could be detected even from the same frame of reference. You would feel that motion as the mass of your body resisted the change in your velocity. If you were in a car that began to move faster or changed direction, your body would be pushed back into the seat or shoved to the left or the right by its inertia. You would know that the vehicle you were traveling in had changed its rate of motion even if you couldn't look outside and see the vehicle change its position relative to other objects. Was accelerated motion some kind of absolute motion? Einstein set himself the task of explaining why the effects of accelerated motion were different from uniform motion and just what was happening when a body accelerated.

Einstein realized that there was one sense in which acceleration was relative. Without an observation from another frame of reference, you could not tell the difference between acceleration and gravity. Einstein imagined a person trapped in a closed, elevator-sized box holding a ball. If that box rested on the surface of the earth, when the person released the ball it would fall toward the bottom of the box at the same rate of acceleration that all objects fall toward the center of the earth, 32 feet, or 9.8 meters, per second per second. Suppose, however, that the box is not resting on the surface of the earth but is located somewhere in space where there was no gravitational field. Suppose then that the box begins accelerating "upward," that is, away from the floor of the box, at 32 feet per second per second. When the person inside the box released the ball, it would fall toward the floor in exactly the same way as if gravity had pulled on it. The person inside the box could not tell if the ball was falling in a gravitational field or if it was falling because of its inertia as the box around it accelerated upward. Einstein called this the principle of equivalence between acceleration and gravity.

A beam of light would also reveal this equivalence, Einstein realized, because, traveling as fast as anything in the universe can travel, light cannot be made to accelerate any faster. Suppose the person in the box held a flashlight instead of a ball and shone that light horizontally on the opposite wall. If the box then accelerated upward very rapidly, the beam of light should be seen to curve downward. It could not keep moving in a straight line because that would mean accelerating along with the box and traveling faster than the speed of light. Here Einstein had the courage to follow his own reasoning through to its logical conclusion. If it was not possible, from the same frame of reference, to tell the difference between acceleration and gravity, then all the effects observed during acceleration ought to be observed in a gravitational field. Therefore, a beam of light had to bend under the influence of gravity.

A beam of light, however, has no mass, so how could gravity act on it to bend its path? After ten years of thinking about this problem, Einstein reached a remarkable conclusion. He concluded that there was no "force" of gravity at all, not in the sense that Newton had understood it. On the contrary: In the presence of a massive object, space itself took on a curvature, the way a cannonball might stretch a rubber sheet it was placed upon. All bodies in motion that encountered such curvature had to move through space in a curved path, the way a golf ball alters its path as it rolls across uneven ground. Gravity was simply a property of the geometry of space, and Einstein's general theory of relativity is really a new theory of gravitation.

Space, however, is empty. Where was the underlying structure that could curve? How could nothing bend? The answer came out of the complex mathematics Einstein had to develop to replace Newton's formulas. These new equations showed that space and time were not independent entities. The universe, in fact, was really a four-dimensional structure of space-time. The three dimensions of space might appear normal, but in the presence of a massive object like a planet, time slowed down. It can be shown mathematically and geometrically that if time slows down near a massive object, observers will see objects move in curved paths. It is space-time, not space, that is curved. The "force" of gravity is an illusion produced by this curvature.

This insight also brought the special and general theories of relativity together in an interesting way. When an object accelerated close to the speed of light, a clock moving with that object would appear to outside observers to slow down. It could be said that all objects existed in space-time, and no object could travel through space-time faster than the speed of light. As an object went faster and faster through space, it had to move more slowly through time, so that its speed through all dimensions could not exceed the speed of light. If an object were able to move at the speed of light through space, its motion through time would have to be zero. An object at complete rest would have a zero velocity through the dimensions of space but would be moving through time at the speed of light. The speed of light, therefore, is the speed of time.

Einstein published his general theory of relativity in 1916. It was not until 1919 that a team of British scientists verified the bending of starlight during a solar eclipse and provided experimental proof for the theory. From that time on, general relativity was accepted as the correct theory, even though few people could grasp its central concepts.

THE EXPANDING UNIVERSE

In 1929, astronomer Edwin Hubble (1889–1953) both confirmed and added a new twist to Einstein's ideas. Before Hubble, all that was known of the universe were the stars in our own galaxy. Hubble began to study the small, fuzzy patches of light in the night sky that were definitely not stars, but which were believed to be clouds of hydrogen gas illuminated by stars within them. Some of these fuzzy patches were clouds of gas, but others Hubble was able to prove were conglomerations of stars just like our galaxy, only very far away.

Astronomers measured great distances in terms of light-years, the distance a wave of light would travel in a year. Since light travels at 300,000 kilometers, or 186,000 miles, per second, the distance it travels in a year is enormous. Hubble calculated that the Andromeda galaxy, one of the nearest to our own Milky Way galaxy, was about 800,000 light-years away. Other galaxies had to be millions of light-years away. Suddenly the entire scale of the universe had to be revised in scientists' minds.

Hubble also analyzed the light from these distant galaxies and found that, if these galaxies were composed of normal stars like our own galaxy, the light from them was redder than it should have been. He attributed this to the Doppler effect, named after

Edwin Hubble discovered that there are many other galaxies in the universe in addition to the one in which we live, the Milky Way.

Hubble found that distant galaxies are moving away from Earth and that the farther away a galaxy is, the faster it seems to be moving away from us. The universe is expanding.

Austrian physicist Christian Doppler (1803–1853), who had discovered that the frequency of sound waves will change if an object making that sound is moving toward or away from you. If the object is moving toward you, the sound waves arrive compressed together, with a shorter wavelength, and you hear a higher pitched sound. If an object is moving away from you, the waves are stretched out, with a lower frequency, and you hear a lower pitched sound. Since red light has a lower frequency than light of other colors, Hubble concluded that the distant galaxies were moving away from Earth. In fact, this "red shift" showed that the farther away a galaxy was, the faster it was moving away. The universe was expanding.

No matter in what direction Hubble looked, the galaxies were receding from Earth. That was odd. Not since the Middle Ages had anyone suggested that Earth was at the center of the universe, and no one wanted to suggest that now, not after Einstein's disproof of absolute motion. Nor was it necessary to believe that Earth was at the center of an expanding universe, thanks to Einstein's theories. Without great mental effort, we can see and conceive of space only as dimensionally expanding, whereas it is really space-time that is expanding. We can't easily conceive of the "shape" of the whole universe. From any point of view within space, all the galaxies will appear to be moving away from each other, like dots placed on the surface of an expanding balloon. We cannot sense more than the surface of this balloon, where we reside in space, although the balloon is really a more complex structure.

Georges Lemaitre pioneered the big bang theory, which indicates the universe has been expanding over billions of years.

Hubble estimated the radius of the universe to be 13 billion light-years. That was not only the radius of the universe; it was the age of the universe, since time expands along with space. Using Hubble's data and extrapolating backward, Belgian astronomer Georges Lemaitre (1894–1966) concluded that at an earlier point in time all the matter in the universe was much closer together. And 13 billion years ago, all matter was compressed into what was variously called a "cosmic egg" or "superatom," from which the present universe was created in a "big bang."

That is where we stand now. We live in an expanding, dynamic universe that is growing larger and older as we speak. It is a universe that has no up or down, no right or left orientation, whose expansion appears the same from any perspective. The only constant in this universe is the speed of light, and all other fundamental physical properties—space, time, mass—are relative. We will measure those properties differently, depending on where in the universe we are and how we are moving in relation to the things we are trying to measure. Newton's laws describing force and acceleration work well enough for most human-scale problems, but they do not really describe the true nature of things, as we discover when we study objects that are moving extremely fast. When objects begin to move extremely fast through space, they begin to move more slowly through time. Motion is a much

According to the big bang theory, the universe began with an instantaneously expanding point of one atom and has continued to expand over billions of years, gradually increasing the distance between our galaxy and others.

more complicated phenomenon than it first appeared to the Greek philosophers.

Will another scientific revolution deepen our understanding of what the universe is? It is impossible to predict such events. Today, cosmologists, those who study the origin and evolution of the universe, are trying to determine if the universe will expand forever, or if, under the influence of gravity, it will eventually contract. Other scientists are trying to discover the relationship of gravity to the other forces of nature. Still others speculate about multiple expanding universes, like a collection of soap bubbles. It certainly seems as if there are plenty of scientific questions left to answer, and the effort to answer them may produce a new understanding of the cosmos. But we have come an extraordinarily long way in our understanding of how objects move. We have reached the point where our scientific understanding even goes beyond our ability to imagine the reality our formulas describe, and that is a considerable achievement.

Glossary

acceleration A change in velocity, with velocity defined as both the speed and direction of motion of a body.

constant A fixed, unchanging value in a mathematical equation.

Copernican system The theory that the planets orbit around the Sun. Copernicus developed the theory in the early sixteenth century. Also known as the heliocentric, or Sun-centered, system.

equivalence Einstein's theory that, without an observer in a different frame of reference, you cannot tell the difference between acceleration and gravity.

friction   The resistance to motion experienced by a body moving through or in contact with another material.

geocentric system   A theory in which Earth is placed at the center of the solar system. Also known as the Ptolemaic system.

gravity   The term used to describe the attraction between two bodies. Newton showed that the force of gravity is directly proportional to the product of the objects' masses and inversely proportional to the square of the distance between them. In other words, the force of attraction between two bodies depends on their masses, and it lessens the farther apart they are.

heliocentric system   The description of planetary motion given by Copernicus that puts the Sun at the center of the solar system.

inertia   Resistance to a change in motion, or the tendency of a body to remain in motion in a

straight line or at rest unless acted on by a
force. All material objects that have mass
have inertia.

mass A measure of the quantity of matter that a
body contains. Technically, this is not the same
as weight, since weight is a measure of the
force of gravity acting on a mass. Scientists
also define mass as the amount of inertia, or
resistance to change in motion, that an object
has. Mass varies with the speed of a body,
increasing as the body approaches the speed
of light. Measuring the quantity of matter in a
body will yield its rest mass, which increases
as the body acclerates.

orbit The path of one celestial body around another,
such as the Moon around Earth, caused by
gravitational attraction.

relativity In physics, the principle that measurements of an object's fundamental properties

will vary depending upon the frame of reference and motion of the observer. Galileo devised the first principle of relativity to account for the fact that Earth could move though no one on Earth could sense its movement. Einstein's theory of special relativity demonstrated that space and time and mass could be perceived differently from different frames of reference.

For More Information

WEB SITES

American Institute of Physics
http://www.aip.org
Extensive links and a good history section.

Astronomy Notes
http://www.astronomynotes.com
Besides general astronomy, this site has good notes on Newton and Einstein.

Galileo Galilei
http://galileo.imss.firenze.it/museo/b/egalilg.html

FOR MORE INFORMATION

The Galileo Project
http://es.rice.edu/ES/humsoc/Galileo

The Internet Encyclopedia of Philosophy
http://www.utm.edu/research/iep
Features the works of Greek philosophers and other early scientists, including Copernicus.

Isaac Newton
http://www.newton.org.uk
British site with extensive links.

Nova (Public Broadcasting System)
http://www.pbs.org/wgbh/nova/einstein
Web pages for the PBS television series *Nova* on Einstein with easy to understand explanations of some of his work.

for further Reading

EASIER READING

Bixby, William. *The Universe of Galileo and Newton.*
New York: American Heritage Publishing Co., 1964.

Bolton, Sarah K. *Famous Men of Science.* 4th ed. New
York: Thomas Y. Crowell Company, 1960.

Dispezio, Michael A. *Awesome Experiments
in Force and Motion.* New York: Sterling
Publications, 1998.

Gardner, Robert. *Famous Experiments You Can Do.*
New York: Franklin Watts, 1990.

Hightower, Paul W. *Galileo: Astronomer and Physicist.*
Springfield, NJ: Enslow Publishers, 1997.

Lafferty, Peter. *Forces and Motion.* Austin, TX: Raintree Steck-Vaughn, 2001.

Severance, John B. *Einstein: Visionary Scientist.* New York: Clarion Books, 1999.

MORE DIFFICULT READING

Barnett, Lincoln. *The Universe and Dr. Einstein.* Alexandria, VA: Time-Life Books, 1982.

Casper, Barry M., and Richard J. Noer. *Revolutions in Physics.* New York: W. W. Norton & Company, 1972.

Cohen, I. Bernard. *The Birth of a New Physics.* Rev. ed. New York: W. W. Norton & Company, 1985.

Drake, Stillman. *Galileo at Work: His Scientific Biography.* New York: Dover Publications, 1995.

Index

Credits

ABOUT THE AUTHOR

Jeremy Roberts has written several books for young readers, including three works in Rosen Publishing's Holocaust Biographies series. He has been in motion all his life, but of course that's only relative.

PHOTO CREDITS

Cover © David Nunuk/Science Photo Library; cover inset © John Burwell/Pictor; p. 12 © VCG/FPG; p. 14 © Gary Adams/Index Stock; p. 22 © Northwind Picture Archives; p. 27 © Ewing Galloway/Index Stock; p. 28 © Jeremy Burgess/Science Photo Library; pp. 18, 21, 30, 32, 41, 46, 58, 88, 92 © Bettmann/Corbis; pp. 44, 60, 70 © Archive Photos; p. 52 © NASA; p. 55 © Hulton Getty/Archive Photos; p. 63 © Baldwin H. Ward & Kathryn

CREDITS

C. Ward/Corbis; p. 90 © W. N. Colley, E. Turner, J. A. Tyson, and NASA; p. 94 © E. Schrempp/Photo Researchers. Diagrams on pp. 37, 64, 72, 81 by Geri Giordano.

DESIGN AND LAYOUT

Evelyn Horovicz